SEVEN BRIEF
LESSONS
ON PHYSICS

SEVEN BRIEF
LESSONS
ON PHYSICS

CARLO ROVELLI

TRANSLATED BY SIMON CARNELL
AND ERICA SEGRE

RIVERHEAD BOOKS
NEW YORK
2016

RIVERHEAD BOOKS
An imprint of Penguin Random House LLC
375 Hudson Street
New York, New York 10014

First American edition published by Riverhead Books 2016
Copyright © 2014 by Carlo Rovelli
Translation copyright © 2015 by Simon Carnell and Erica Segre
Originally published in Italian under the title *Sette brevi lezioni di fisica* by Adelphi Edizioni, Milan
English translation published in Great Britain by Allen Lane, an imprint of Penguin Random House UK

ISBN 978-0-399-18441-3

Printed in the United States of America
10   9   8   7   6   5   4   3   2   1

Book design by Lauren Kolm

CONTENTS

PREFACE

These lessons were written for those who know little or nothing about modern science. Together they provide a rapid overview of the most fascinating aspects of the great revolution that has occurred in physics in the twentieth and twenty-first centuries, and of the questions and mysteries that this revolution has opened up. Because science shows us how to better understand the world, but it also reveals to us just how vast is the extent of what is still not known.

The first lesson is dedicated to Albert Einstein's general theory of relativity, the "most beautiful of theories."

The second to quantum mechanics, where the most baffling aspects of modern physics lurk. The third is dedicated to the cosmos: the architecture of the universe that we inhabit; the fourth to its elementary particles. The fifth deals with quantum gravity: the attempts that are under way to construct a synthesis of the major discoveries of the twentieth century. The sixth is on probability and the heat of black holes. The final section of the book returns to ourselves and asks how it is possible to think about our existence in the light of the strange world described by physics.

The lessons are expansions of a series of articles published in the Sunday supplement of the Italian newspaper *Il Sole 24 Ore*. I would like to thank in particular Armando Massarenti, who can be credited with opening up the cultural pages of a Sunday paper to science and allowing light to be thrown on the role of this integral and vital aspect of our culture.

FIRST LESSON

The Most Beautiful of Theories

In his youth Albert Einstein spent a year loafing aimlessly. You don't get anywhere by not "wasting" time—something, unfortunately, that the parents of teenagers tend frequently to forget. He was in Pavia. He had joined his family, having abandoned his studies in Germany, unable to endure the rigors of his high school there. It was the beginning of the twentieth century, and in Italy the beginning of its industrial revolution. His father, an engineer, was installing the first electricity-generating power plants in the Paduan plains. Albert was reading Kant and attending occasional lectures at the University of Pavia: for pleasure, without being registered there or having to think about exams. It is thus that serious scientists are made.

After this he registered at the University of Zurich and immersed himself in the study of physics. A few years later, in 1905, he sent three articles to the most prestigious scientific journal of the period, the *Annalen der Physik*. Each of these is worthy of a Nobel Prize. The first shows that atoms really exist. The second lays the first foundation for quantum mechanics, which I will discuss in the next lesson. The third presents his first theory of relativity (known today as "special relativity"), the theory that elucidates how time does not pass identically for everyone: two identical twins find that they are different in age if one of them has traveled at speed.

Einstein became a renowned scientist overnight and received offers of employment from various universities. But something disturbed him: despite its immediate acclaim, his theory of relativity does not fit with what we know about gravity, namely, with how things fall. He came to realize this when writing an article summarizing his theory and began to wonder if the law of "universal gravity" as formulated by the father of physics himself, Isaac Newton, was in need of revision in order to make it compatible with the new concept of relativity. He immersed himself in the problem. It would take ten years to resolve. Ten years of frenzied studies, attempts, errors, confusion, mistaken articles, brilliant ideas, misconceived ideas.

Finally, in November 1915, he committed to print an article giving the complete solution: a new theory of gravity, which he called "The General Theory of Relativity," his masterpiece and the "most beautiful of theories," according to the great Russian physicist Lev Landau.

There are absolute masterpieces that move us intensely: Mozart's *Requiem*, Homer's *Odyssey*, the Sistine Chapel, *King Lear*. To fully appreciate their brilliance may require a long apprenticeship, but the reward is sheer beauty—and not only this, but the opening of our eyes to a new perspective upon the world. Einstein's jewel, the general theory of relativity, is a masterpiece of this order.

I remember the excitement I felt when I began to understand something about it. It was summer. I was on a beach at Condofuri in Calabria, immersed in the sunshine of the Hellenic Mediterranean, and in the last year of my university studies. Undistracted by schooling, one studies best during vacations. I was studying with the help of a book that had been gnawed at the edges by mice because at night I'd used it to block the holes of these poor creatures in the rather dilapidated, hippie-ish house on an Umbrian hillside where I used to take refuge from the tedium of university classes in Bologna. Every so often I would raise my eyes from the book and look at the glittering sea: it seemed to me that I was actually seeing the curvature of space and time imagined by Einstein. As if by magic: as if a friend were whispering into my ear an extraordinary hidden truth, suddenly raising the veil of reality to disclose a simpler, deeper order. Ever since we discovered that Earth is round and turns like a mad spinning-top, we have understood that reality is not as it appears to us: every time we glimpse a new aspect of it, it is a deeply emotional experience. Another veil has fallen.

But among the numerous leaps forward in our understanding that have succeeded one another over the course of history, Einstein's is perhaps unequaled. Why?

In the first place because, once you understand how it works, the theory has a breathtaking simplicity. I'll summarize the idea.

Newton had tried to explain the reason why things fall and the planets turn. He had imagined the existence of a "force" that draws all material bodies toward one another and called it "the force of gravity." How this force was exerted between things distant from each other, without there being anything between them, was unknown—and the great father of modern science was cautious of offering a hypothesis. Newton had also imagined that bodies move through space and that space is a great empty container, a large box that enclosed the universe, an immense structure through which all objects run true until a force obliges their trajectory to curve. What this "space" was made of, this container of the world he invented, Newton could not say. But a few years before the birth of Einstein two great British physicists, Michael Faraday and James Maxwell, had added a key ingredient to Newton's cold world: the electromagnetic field. This field is a real entity that, diffused everywhere, carries radio waves, fills space, can vibrate and oscillate like the surface of a lake, and "transports" the electrical force. Since his youth Einstein had been fascinated by this electromagnetic field that turned the rotors in the power stations built by his father, and he soon came to understand that gravity, like electricity, must be conveyed by a field as well: a "gravitational field" analogous to the "electrical field" must exist. He aimed at understanding how this "gravitational field" worked and how it could be described with equations.

And it is at this point that an extraordinary idea occurred to him, a stroke of pure genius: the gravitational field is not *diffused through* space; the gravitational field *is that space* itself. This is the idea of the general theory of relativity. Newton's "space," through which things move, and the "gravitational field" are one and the same thing.

It's a moment of enlightenment. A momentous simplification of the world: space is no longer something distinct from matter—it is one of the "material" components of the world. An entity that undulates, flexes, curves, twists. We are not contained within an invisible, rigid infrastructure: we are immersed in a gigantic, flexible snail shell. The sun bends space around itself, and Earth does not turn around it because of a mysterious force but because it is racing directly in a space that inclines, like a marble that rolls in a funnel. There are no mysterious forces generated at the center of the funnel; it is the curved nature of the walls that causes the marble to roll. Planets circle around the sun, and things fall, because space curves.

How can we describe this curvature of space? The

most outstanding mathematician of the nineteenth century, Carl Friedrich Gauss, the so-called prince of mathematicians, had written mathematical formulas to describe two-dimensional curvilinear surfaces, such as the surfaces of hills. Then he had asked a gifted student of his to generalize the theory to encompass spaces in three or more dimensions. The student in question, Bernhard Riemann, had produced an impressive doctoral thesis of the kind that seems completely useless. The result of Riemann's thesis was that the properties of a curved space are captured by a particular mathematical object, which we know today as Riemann's curvature and indicate with the letter R. Einstein wrote an equation that says that R is equivalent to the energy of matter. That is to say: space curves where there is matter. That is it. The equation fits into half a line, and there is nothing more. A vision—that space curves—became an equation.

But within this equation there is a teeming universe. And here the magical richness of the theory opens up into a phantasmagorical succession of predictions that resemble the delirious ravings of a madman but have all turned out to be true.

To begin with, the equation describes how space bends around a star. Due to this curvature, not only do planets orbit around the star but light stops moving in a straight line and deviates. Einstein predicted that the sun causes light to deviate. In 1919 this deviance was measured and the prediction verified. But it isn't only space that curves; time does too. Einstein predicted that time passes more quickly high up than below, nearer to Earth. This was measured and turned out to be the case. If a person who has lived at sea level meets up with his twin who has lived in the mountains, he will find that his sibling is slightly older than he. And this is just the beginning.

When a large star has burned up all of its combustible substance (hydrogen), it goes out. What remains is no longer supported by the heat of the combustion and collapses under its own weight, to a point where it bends space to such a degree that it plummets into an actual hole. These are the famous "black holes." When I was studying at university they were considered to be the barely credible predictions of an esoteric theory. Today they are observed in the sky in their hundreds and are studied in great detail by astronomers.

But this is still not all. The whole of space can expand and contract. Furthermore, Einstein's equation shows that space cannot stand still; it *must be* expanding. In 1930 the expansion of the universe was actually observed. The same equation predicts that the expansion ought to have been triggered by the explosion of a young, extremely small, and extremely hot universe: by what we now know as the "big bang." Once again, no one believed this at first, but the proof mounted up until "cosmic background radiation"—the diffuse glare that remains from the heat generated by the original explosion—was actually observed in the sky. The prediction arising from Einstein's equation turned out to be correct. And further still, the theory contends that space moves like the surface of the sea. The effects of these "gravitational waves" are observed in the sky on binary stars and correspond to the predictions of the theory even to the astonishing precision of one part to one hundred billion. And so forth.

In short, the theory describes a colorful and amazing world where universes explode, space collapses into bottomless holes, time sags and slows near a planet, and the unbounded extensions of interstellar space ripple and sway like the surface of the sea . . . And all of this, which emerged gradually from my mice-gnawed book, was not a tale told by an idiot in a fit of lunacy or a hallucination caused by Calabria's burning Mediterranean sun and its dazzling sea. It was reality.

Or better, a glimpse of reality, a little less veiled than

our blurred and banal everyday view of it. A reality that seems to be made of the same stuff that our dreams are made of, but that is nevertheless more real than our clouded, quotidian dreaming.

All of this is the result of an elementary intuition: that space and gravitational field are the same thing. And of a simple equation that I cannot resist giving here, even though you will almost certainly not be able to decipher it. Perhaps anyone reading this will still be able to appreciate its wonderful simplicity:

$$R_{ab} - \tfrac{1}{2} R\, g_{ab} = T_{ab}$$

That's it.

You would need, of course, to study and digest Riemann's mathematics in order to master the technique to read and use this equation. It takes a little commitment and effort. But less than is necessary to come to appreciate the rarefied beauty of a late Beethoven string quartet. In both cases the reward is sheer beauty and new eyes with which to see the world.

The two pillars of twentieth-century physics—general relativity, of which I spoke in the first lesson, and quantum mechanics, which I'm dealing with here—could not be more different from each other. Both theories teach us that the fine structure of nature is more subtle than it appears. But general relativity is a compact gem: conceived by a single mind, that of Albert Einstein, it's a simple and coherent vision of gravity, space, and time. Quantum mechanics, or "quantum theory," on the other hand, has gained unequaled experimental success and led to applications that have transformed our everyday lives (the computer on which I write, for example), but more than a century after its birth it remains shrouded in mystery and incomprehensibility.

It's said that quantum mechanics was born precisely in the year 1900, virtually ushering in a century of intense thought. The German physicist Max Planck calculated the electric field in equilibrium in a hot box. To do this he used a trick: he imagined that the energy of the field is distributed in "quanta," that is, in packets or lumps of energy. The procedure led to a result that perfectly reproduced what was measured (and therefore must be in some fashion correct) but clashed with everything that was known at the time. Energy was considered to be something that varied continuously, and there was no reason to treat it as if it were made up of small building blocks. To treat energy as if it were made up of finished packages had been, for Planck, a peculiar trick of calculation, and he did not himself fully understand the reason for its effectiveness. It was to be Einstein once again who, five years later, came to understand that the "packets of energy" were real.

Einstein showed that light is made of packets: particles of light. Today we call these "photons." He wrote, in the introduction to his article:

It seems to me that the observations associated with blackbody radiation, fluorescence, the production of cathode rays by ultraviolet light, and other related phenomena connected with the emission or transformation of light are more readily understood if one assumes that the energy of light is discontinuously distributed in space. In accordance with the assumption to be considered here, the energy of a light ray spreading out from a point source is not continuously distributed over an increasing space but consists of a finite number of "energy quanta" which are localized at points in space, which move without dividing, and which can only be produced and absorbed as complete units.

These simple and clear lines are the real birth certificate of quantum theory. Note the wonderful initial "It seems to me . . . ," which recalls the "I think . . ." with which Darwin introduces in his notebooks the great idea that species evolve, or the "hesitation" spoken of by Faraday when introducing for the first time the revolutionary idea of magnetic fields. Genius hesitates.

The work of Einstein is initially treated by colleagues

as the nonsensical juvenilia of an exceptionally brilliant youth. Subsequently it will be for the same work that he is awarded the Nobel Prize. If Planck is the father of the theory, Einstein is the parent who nurtured it.

But like all offspring, the theory then went its own way, unrecognized by Einstein himself. In the second and third decades of the twentieth century it was the Dane Niels Bohr who pioneered its development. It was Bohr who understood that the energy of electrons in atoms can take on only certain values, like the energy of light, and crucially that electrons can only "jump" between one atomic orbit and another with determined energies, emitting or absorbing a photon when they jump. These are the famous "quantum leaps." And it was in his institute in Copenhagen that the most brilliant young minds of the century gathered together to investigate and try to bring order to these baffling aspects of behavior in the atomic world, and to build from it a coherent theory. In 1925 the equations of the theory finally appeared, replacing the entire mechanics of Newton.

It's difficult to imagine a greater achievement. At one stroke, everything makes sense, and you can calculate everything. Take one example: do you remember the periodic table of elements, devised by Dmitri Mendeleev, which lists all the possible elementary substances of which the universe is made, from hydrogen to uranium, and which was hung on so many classroom walls? Why are precisely these elements listed there, and why does the periodic table have this particular structure, with these periods, and with the elements having these specific properties? The answer is that each element corresponds to one solution of the main equation of quantum mechanics. The whole of chemistry emerges from a single equation.

The first to write the equations of the new theory, basing them on dizzying ideas, would be a young German of genius, Werner Heisenberg.

Heisenberg imagined that electrons do not *always* exist. They only exist when someone or something watches them, or better, when they are interacting with something else. They materialize in a place, with a calculable probability, when colliding with something else. The "quantum leaps" from one orbit to another are the only means they have of being "real": an electron is a set of jumps from one interaction to another. When nothing disturbs it, it is not in any precise place. It is not in a "place" at all.

It's as if God had not designed reality with a line that was heavily scored but just dotted it with a faint outline.

In quantum mechanics no object has a definite position, except when colliding headlong with something else. In order to describe it in mid-flight, between one interaction and another, we use an abstract mathematical formula that has no existence in real space, only in abstract mathematical space. But there's worse to come: these interactive leaps with which each object passes from one place to another do not occur in a predictable way but largely at random. It is not possible to predict where an electron will reappear but only to calculate the *probability* that it will pop up here or there. The question of probability goes to the heart of physics, where everything had seemed to be regulated by firm laws that were universal and irrevocable.

Does it seem absurd? It also seemed absurd to Einstein. On the one hand he proposed Heisenberg for the Nobel Prize, recognizing that he had understood something fundamental about the world, while on the other he didn't miss any occasion to grumble that this did not make much sense.

The young lions of the Copenhagen group were dismayed: how was it possible that *Einstein* should think this? Their spiritual father, the man who had shown the courage to think the unthinkable, now retreated and was afraid of this new leap into the unknown that he himself had triggered. The same Einstein who had shown that time is not universal and that space is curved was now saying that the world cannot be *this* strange.

Patiently, Bohr explained the new ideas to Einstein. Einstein objected. He devised mental experiments to show that the new ideas were contradictory: "Imagine a box filled with light, from which we allow a single photon to escape for an instant . . ." So begins one of his famous examples, the mental experiment of the "box of light." In the end Bohr always managed to find an answer with which to rebut these objections. For years, their dialogue continued by way of lectures, letters, articles . . . During the course of the exchange both great men needed to backtrack, to change their thinking. Einstein had to admit that there was actually no contradiction within the new ideas. Bohr had to recognize that things were not as simple and clear as he'd initially thought. Einstein did not want to relent on what was for him the key issue: that there was an objective reality independent of whoever interacts with whatever. Bohr would not relent on the validity of the profoundly new way in which the real was conceptualized by the new theory. Ultimately, Einstein conceded that the theory was a giant step forward in our understanding of the world, but he remained convinced that things could not be as strange as it proposed—that "behind" it there must be a further, more reasonable explanation.

A century later we are at the same point. The equations of quantum mechanics and their consequences are used daily in widely varying fields—by physicists, engineers, chemists, and biologists. They are extremely useful in all contemporary technology. Without quantum mechanics there would be no transistors. But they remain mysterious. For they do not describe what happens to a physical system but only how a physical system affects another physical system.

What does this mean? That the essential reality of a system is indescribable? Does it mean that we lack only a piece of the puzzle? Or does it mean, as it seems to me, that we must accept the idea that reality is only interaction? Our knowledge grows, in real terms. It allows us to do new things that we had previously not even imagined. But that growth has opened up new questions. New mysteries. Those who use the equations of the theory in the laboratory carry on regardless, but in articles and conferences that have been increasingly numerous in recent years, physicists and philosophers continue to search. What is quantum theory a century after its birth? An extraordinary dive deep into the nature of reality? A blunder that works, by chance? Part of an incomplete puzzle? Or a clue to something profound regarding the structure of the world that we have not yet properly digested?

When Einstein died, his greatest rival, Bohr, found for him words of moving admiration. When a few years later Bohr in turn died, someone took a photograph of the blackboard in his study. There's a drawing on it. A drawing of the "light-filled box" in Einstein's thought experiment. To the very last, the desire to challenge oneself and understand more. And to the very last: doubt.

THIRD LESSON

The Architecture of the Cosmos

In the first half of the twentieth century Einstein described the workings of space and time, while Niels Bohr and his young disciples captured in equations the strange quantum nature of matter. In the second half of the century physicists built upon these foundations, applying the two new theories to widely varying domains of nature: from the macrocosmic structure of the universe to the microcosm of elementary particles. I speak of the first of these in this lesson, and of the second in the next.

This lesson is made up mostly of simple drawings.

The reason for this is that before experiments, measurements, mathematics, and rigorous deductions, science is above all about visions. Science begins with a vision. Scientific thought is fed by the capacity to "see" things differently than they have previously been seen. I want to offer here a brief, modest outline of a journey between visions.

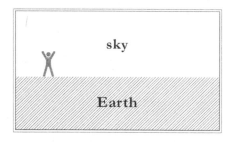

This first image represents how the cosmos was conceptualized for millennia: Earth below, the sky above. The first great scientific revolution, accomplished by Anaximander twenty-six centuries ago when trying to figure out how it is possible that the sun, moon, and stars revolve around us, replaced the above image of the cosmos with the one at the top of the next page.

Now the sky is all around Earth, not just above it, and Earth is a great stone that floats suspended in space, without falling. Soon someone (perhaps Parmenides, perhaps

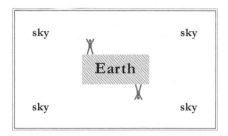

Pythagoras) realized that the sphere is the most reasonable shape for this flying Earth for which all directions are equal—and Aristotle devised convincing scientific arguments to confirm the spherical nature of both Earth and the heavens around it where celestial objects run their course. Here is the resultant image of the cosmos:

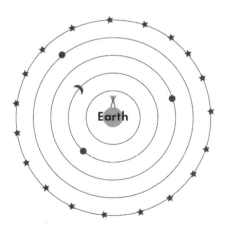

Carlo Rovelli

And this cosmos, as described by Aristotle in his book *On the Heavens*, is the image of the world that remained characteristic of Mediterranean civilizations right up until the end of the Middle Ages. It's the image of the world that Dante and Shakespeare studied at school.

The next leap was accomplished by Copernicus, inaugurating what has come to be called the great scientific revolution. The world for Copernicus is not so very different from Aristotle's:

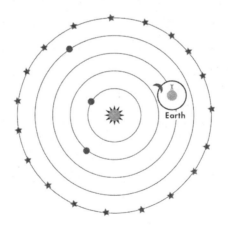

But there is in fact a key difference. Taking up an idea already considered in antiquity, Copernicus understood and showed that our Earth is not at the center of the dance of the planets but that the sun is there instead. Our planet becomes one among the others, turning at high speed upon its axis and around the sun.

The growth of our knowledge continued, and with improved instruments it was soon learned that the solar system itself is only one among a vast number of others, and that the sun is no more than a star like others. An infinitesimal speck in a vast cloud of one hundred billion stars—the Galaxy:

In the 1930s, however, precise measurements by astronomers of the nebulae—small whitish clouds between the stars—showed that the Galaxy itself is a speck of dust in a huge cloud of galaxies, which extends as far as the eye can see using even our most powerful telescopes. The world has now become a uniform and boundless expanse.

The illustration below is not a drawing; it's a photograph taken by the Hubble telescope in orbit, showing a deeper image of the sky than any seen previously with the most powerful of our telescopes; seen with the naked eye, it would be a minute piece of extremely black sky. Through the Hubble telescope a dusting of vastly distant dots appears. Each black dot in the image is a galaxy containing a hundred billion suns similar to ours. In the past few years it has been observed that the majority of these suns are orbited by planets. There are therefore in the universe thousands of billions of billions of billions of planets such as Earth. And in every direction in which we look, this it what appears:

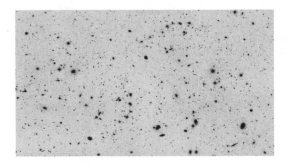

But this endless uniformity, in turn, is not what it seems. As I explained in the first lesson, space is not flat but curved. We have to imagine the texture of the universe, with its splashes of galaxies, being moved by waves similar to those of the sea, sometimes so agitated as to create the gaps that are black holes. So let's return to a drawn image, in order to represent this universe furrowed by great waves:

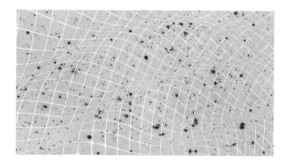

And finally, we now know that this immense, elastic cosmos, studded with galaxies and fifteen billion years in the making, emerged from an extremely hot and dense small cloud. To represent this vision, we no longer need to draw the universe but to draw its entire history. Here it is, diagrammatically:

Carlo Rovelli

The universe began as a small ball and then exploded to its present cosmic dimensions. This is our current image of the universe, on the grandest scale that we know.

Is there anything else? Was there something before? Perhaps, yes. I'll talk about it after a couple of lessons. Do other similar universes exist, or different ones? We do not know.

Within the universe described in the previous lesson, light and things move. Light is made up of photons, the particles of light intuited by Einstein. The things we see are made of atoms. Every atom consists of a nucleus surrounded by electrons. Every nucleus consists of tightly packed protons and neutrons. Both protons and neutrons are made up of even smaller particles that the American physicist Murray Gell-Mann named "quarks," inspired by a seemingly nonsensical word in a nonsensical phrase in James Joyce's *Finnegans Wake*: "Three quarks for Muster Mark!" Everything we touch is therefore made of electrons and of these quarks.

The force that "glues" quarks inside protons and neutrons is generated by particles that physicists, with little sense of the ridiculous, call "gluons."

Electrons, quarks, photons, and gluons are the components of everything that sways in the space around us. They are the "elementary particles" studied in particle physics. To these particles a few others are added, such as the neutrinos, which swarm throughout the universe but have little interaction with us, and the "Higgs bosons," recently detected in Geneva in CERN's Large Hadron Collider. But there are not many of these, fewer than ten types, in fact. A handful of elementary ingredients that act like bricks in a gigantic Lego set, and with which the entire material reality surrounding us is constructed.

The nature of these particles, and the way they move, is described by quantum mechanics. These particles do not have a pebble-like reality but are rather the "quanta" of corresponding fields, just as photons are the "quanta" of the electromagnetic field. They are elementary excitations of a moving substratum similar to the field of Faraday and Maxwell. Minuscule moving wavelets. They disappear and reappear according to the strange laws of quantum mechanics, where everything that exists is never stable and is nothing but a jump from one interaction to another.

Even if we observe a small, empty region of space in which there are no atoms, we still detect a minute swarming of these particles. There is no such thing as a real void, one that is completely empty. Just as the calmest sea looked at closely sways and trembles, however slightly, so the fields that form the world are subject to minute fluctuations, and it is possible to imagine its basic particles having brief and ephemeral existences, continually created and destroyed by these movements.

This is the world described by quantum mechanics and particle theory. We have arrived very far from the mechanical world of Newton, where minute, cold stones eternally wandered on long, precise trajectories in geometrically immutable space. Quantum mechanics and experiments with particles have taught us that the world is a continuous, restless swarming of things, a continuous coming to light and disappearance of ephemeral entities. A set of vibrations, as in the switched-on hippie world of the 1960s. A world of happenings, not of things.

The details of particle theory were built gradually in the 1950s, 1960s, and 1970s by some of the century's greatest physicists, such as Richard Feynman and Gell-

Mann. This work of construction led to an intricate theory, based on quantum mechanics and bearing the not very romantic title of "the Standard Model of elementary particles." The Standard Model was finalized in the 1970s, after a long series of experiments that confirmed all predictions. Its final confirmation occurred in 2013 with the discovery of the Higgs boson.

But despite the long series of successful experiments, the Standard Model has never been taken entirely seriously by physicists. It's a theory that looks, at least at first sight, piecemeal and patched together. It's made up of various pieces and equations assembled without clear order. A certain number of fields (but why *these*, exactly?) interacting among themselves with certain forces (but why *these* forces?) each determined by certain constants (but why precisely *these* values?) showing certain symmetries (but again, why *these*?). We're far from the simplicity of the equations of general relativity and of quantum mechanics.

The very way in which the equations of the Standard Model make predictions about the world is also absurdly convoluted. Used directly, these equations lead to nonsensical predictions where each calculated quantity turns out to be infinitely large. To get meaningful results, it is necessary to imagine that the parameters entering into them are themselves infinitely large, in order to counterbalance the absurd results and make them reasonable. This convoluted and baroque procedure is given the technical term "renormalization." It works in practice but leaves a bitter taste in the mouth of anyone desiring simplicity of nature. In the last years of his life, the greatest scientist of the twentieth century after Einstein, Paul Dirac, the great architect of quantum mechanics and author of the first and principal equations of the Standard Model, repeatedly expressed his dissatisfaction at this state of things, concluding that "we have not yet solved the problem."

In addition, a striking limitation of the Standard Model has appeared in recent years. Around every galaxy, astronomers observe a large cloud of material that reveals its existence via the gravitational pull that it exerts upon stars and by the way it deflects light. But this great cloud, of which we observe the gravitational effects, cannot be seen directly and we do not know what it is made of. Numerous hypotheses have been proposed, none of which seem to work. It's clear that there is *something* there, but we don't know what. Nowadays it is called "dark matter." Evidence indicates that it is something *not* described by the Standard Model; otherwise we would see it. Something other than atoms, neutrinos, or photons . . .

It is hardly surprising that there are more things in heaven and earth, dear reader, than have been dreamed of in our philosophy—or in our physics. We did not even suspect the existence of radio waves and neutrinos, which fill the universe, until recently. The Standard Model remains the best that we have when speaking today about the world of things; its predictions have all been confirmed, and, apart from dark matter—and gravity as described in the general theory of relativity as the curvature of space-time—it describes well every aspect of the perceived world.

Alternative theories have been proposed, only to be demolished by experiments. A fine theory proposed in the 1970s and given the technical name SU(5), for example, replaced the disordered equations of the Standard Model with a much simpler and more elegant structure. The theory predicted that a proton could disintegrate, with a certain probability, transforming into electrons and quarks. Large machines were constructed to observe protons disintegrating. Physicists dedicated their lives to the search for an observable proton disintegration. (You do not look at one proton at a time, because it takes too long to disintegrate. You take tons of water and surround it with sensitive detectors to observe the effects of disintegration.) But, alas, no proton was ever seen disintegrating. The beautiful theory, SU(5), despite its considerable elegance, was not to the good Lord's liking.

The story is probably repeating itself now with a group of theories known as "supersymmetric," which predicts the existence of a new class of particles. Throughout my career I have listened to colleagues awaiting with complete confidence the imminent appearance of these particles. Days, months, years, and decades have passed—but the supersymmetric particles have not yet manifested themselves. Physics is not only a history of successes.

So, for the moment we have to stay with the Standard Model. It may not be very elegant, but it works remarkably well at describing the world around us. And who knows? Perhaps on closer inspection it is not the model that lacks elegance. Perhaps it is we who have not yet learned to look at it from just the right point of view, one that would reveal its hidden simplicity. For now, this is what we know of matter:

A handful of types of elementary particles, which vibrate and fluctuate constantly between existence and nonexistence and swarm in space, even when it seems that there is nothing there, combine together to infinity like the letters of a cosmic alphabet to tell the immense history of galaxies; of the innumerable stars; of sunlight; of mountains, woods, and fields of grain; of the smiling faces of the young at parties; and of the night sky studded with stars.

FIFTH LESSON

Grains of Space

Despite certain obscurities, infelicities, and still unanswered questions, the physics I have outlined provide a better description of the world than we have ever had in the past. So we should be quite satisfied. But we are not.

There's a paradox at the heart of our understanding of the physical world. The twentieth century gave us the two gems of which I have spoken: general relativity and quantum mechanics. From the first cosmology developed, as well as astrophysics, the study of gravitational waves, of black holes, and much else besides. The second provided the foundation for atomic physics, nuclear physics, the physics of elementary particles, the physics of condensed matter, and much, much more. Two theories, profligate in their gifts, which are fundamental to today's technology and have transformed the way we live. And yet the two theories cannot both be right, at least in their current forms, because they contradict each other.

A university student attending lectures on general relativity in the morning and others on quantum mechanics in the afternoon might be forgiven for concluding that his professors are fools or have neglected to communicate with one another for at least a century. In the morning the world is curved space where everything is continuous; in the afternoon it is a flat space where quanta of energy leap.

The paradox is that both theories work remarkably well. Nature is behaving with us like that elderly rabbi to whom two men went in order to settle a dispute. Having listened to the first, the rabbi says: "You are in the right." The second insists on being heard. The rabbi listens to him and says: "You're also right." Having overheard from the next room, the rabbi's wife then calls out, "But they can't *both* be in the right!" The rabbi reflects and nods before concluding: "And you're right too."

A group of theoretical physicists scattered across the five continents is laboriously trying to settle the issue. Their field of study is called "quantum gravity": its objective is to find a theory, that is, a set of equations—but above all a coherent vision of the world—with which to resolve the current schizophrenia.

It is not the first time that physics finds itself faced with two highly successful but apparently contradictory theories. The effort to synthesize has in the past been rewarded with great strides forward in our understanding of the world. Newton discovered universal gravity by combining Galileo's parabolas with the ellipses of Kepler. Maxwell found the equations of electromagnetism by combining the theories of electricity and of magnetism. Einstein discovered relativity by way of resolving an apparent conflict between electromagnetism and mechanics. A physicist is only too happy when he finds a conflict of this kind between successful theories: it's an extraordinary opportunity. Can we build a conceptual framework for thinking about the world that is compatible with what we have learned about it from *both* theories?

Here, in the vanguard, beyond the borders of knowledge, science becomes even more beautiful—incandescent in the forge of nascent ideas, of intuitions, of attempts. Of roads taken and then abandoned, of enthusiasms. In the effort to imagine what has not yet been imagined.

Twenty years ago the fog was thick. Today paths have appeared that have elicited enthusiasm and optimism. There are more than one of these, so it can't be said that the problem has been resolved. The multiplicity generates controversy, but the debate is healthy: until the fog has lifted completely, it's good to have criticism and opposing views. One of the principal attempts to solve the problem is a direction of research called "loop quantum gravity," pursued by a loose band of researchers working in many countries.

Loop quantum gravity is an endeavor to combine general relativity and quantum mechanics. It is a cautious attempt because it uses only hypotheses already contained within these theories, suitably rewritten to make them compatible. But its consequences are radical: a further profound modification of the way we look at the structure of reality.

The idea is simple. General relativity has taught us that space is not an inert box but rather something dynamic: a kind of immense, mobile snail shell in which we are contained—one that can be compressed and twisted. Quantum mechanics, on the other hand, has taught us that every field of this kind is "made of quanta" and has a fine, granular structure. It immediately follows that physical space is also "made of quanta."

The central result of loop quantum gravity is indeed that space is not continuous, that it is not infinitely divisible but made up of grains, or "atoms of space." These are extremely minute: a billion billion times smaller than the smallest atomic nuclei. The theory describes these "atoms of space" in mathematical form and provides equations that determine their evolution. They are called "loops," or rings, because they are linked to one another, forming a network of relations that weaves the texture of space, like the rings of a finely woven, immense chain mail.

Where are these quanta of space? Nowhere. They are not in space because they are themselves the space. Space is created by the linking of these individual quanta of gravity. Once again, the world seems to be less about objects than about interactive relationships.

But it's the second consequence of the theory that is the most extreme. Just as the idea of a continuous space that contains things disappears, so the idea of an elementary and primal "time" flowing regardless of things also vanishes. The equations describing grains of space and matter no longer contain the variable "time." This doesn't mean that everything is stationary and unchanging. On the contrary, it means that change is ubiquitous—but elementary processes cannot be ordered in a common succession of "instants." At the minute scale of the grains of space, the dance of nature does not take place to the rhythm of the baton of a single orchestral conductor, at a single tempo: every process dances independently with its neighbors, to its own rhythm. The passage of time is internal to the world, is born in the world itself in the relationship between quantum events that comprise the world and are themselves the source of time.

The world described by the theory is thus further distanced from the one with which we are familiar. There is no longer space that "contains" the world, and there is no longer time "in which" events occur. There are only elementary processes wherein quanta of space and matter continually interact with one another. The illusion of space and time that continues around us is a blurred vision of this swarming of elementary processes, just as a calm, clear Alpine lake consists in reality of a rapid dance of myriads of minuscule water molecules.

Viewed in extreme close-up through an ultrapowerful magnifying glass, the penultimate image in our third lesson should show the granular structure of space:

Is it possible to verify this theory experimentally? We are thinking, and trying, but there is as yet no experimental verification. There are, however, a number of different attempts.

One of these derives from the study of black holes. In the heavens we can now observe black holes formed by collapsed stars. Crushed by its own weight, the matter of these stars has collapsed upon itself and disappeared from our view. But where has it gone? If the theory of loop quantum gravity is correct, matter cannot really have collapsed to an infinitesimal point. Because infinitesimal points do not exist—only finite chunks of space. Collapsing under its own weight, matter must have become increasingly dense, up to the point where quantum mechanics must have exerted a contrary, counterbalancing pressure.

This hypothetical final stage in the life of a star, where the quantum fluctuations of space-time balance the weight of matter, is what is known as a "Planck star." If the sun were to stop burning and to form a black hole, it would measure about one and a half kilometers in diameter. Inside this black hole the sun's matter would continue to collapse, eventually becoming such a Planck star. Its dimensions would then be similar to those of an atom. The entire matter of the sun condensed into the space of an atom: a Planck star should be constituted by this extreme state of matter.

A Planck star is not stable: once compressed to the maximum, it rebounds and begins to expand again. This leads to an explosion of the black hole. This process, as seen by a hypothetical observer sitting in the black hole on the Planck star, would be a rebound occurring at great speed. But time does not pass at the same speed for him as for those outside the black hole, for the same reason that in the mountains time passes faster than at sea level. Except that for him, because of the extreme conditions, the difference in the passage of time is enormous, and what for the observer on the star would seem an extremely rapid bounce would appear, seen from outside it, to take place over a very long time. This is why we observe black holes remaining the same for long periods of time: a black hole is a rebounding star seen in extreme slow motion.

It is possible that in the furnace of the first instants of the universe black holes were formed and that some of these are now exploding. If that were true, we could perhaps observe the signals that they emit when exploding, in the form of high-energy cosmic rays coming from the sky, thereby allowing us to observe and measure a direct effect of a phenomenon governed by quantum gravity. It's a bold idea—it might not work, for example, if in the primordial universe not enough black holes were formed to allow us to detect their explosions today. But the search for signals has begun. We shall see.

Another of the consequences of the theory, and one of the most spectacular, concerns the origins of the universe. We know how to reconstruct the history of our world back to an initial period when it was tiny in size. But what about before that? Well, the equations of loop theory allow us to go even further back in the reconstruction of that history.

What we find is that when the universe is extremely compressed, quantum theory generates a repulsive force, with the result that the great explosion, or "big bang," may have actually been a "big bounce." Our world may have actually been born from a preceding universe that contracted under its own weight until it was squeezed into a tiny space before "bouncing" out and beginning to re-expand, thus becoming the expanding universe that we observe around us.

The moment of this bounce, when the universe was contracted into a nutshell, is the true realm of quantum gravity: time and space have disappeared altogether, and the world has dissolved into a swarming cloud of probability that the equations can, however, still describe. And the final image of the third lesson is transformed thus:

Our universe may have been born from a bounce in a prior
phase, passing through an intermediate phase in which
there was neither space nor time.

Physics opens windows through which we see far into the distance. What we see does not cease to astonish us. We realize that we are full of prejudices and that our intuitive image of the world is partial, parochial, inadequate. Earth is not flat; it is not stationary. The world continues to change before our eyes as we gradually see it more extensively and more clearly. If we try to put together what we have learned in the twentieth century about the physical world, the clues point toward something profoundly different from our instinctive understanding of matter, space, and time. Loop quantum gravity is an attempt to decipher these clues and to look a little farther into the distance.

Probability, Time, and the
Heat of Black Holes

Along with the great theories that I have already discussed and that describe the elementary constituents of the world, there is another great bastion of physics that is somewhat different from the others. A single question unexpectedly gave rise to it: "What is heat?"

Until the mid-nineteenth century, physicists attempted to understand heat by thinking that it was a kind of fluid, called "caloric," or two fluids, one hot and one cold. The idea turned out to be wrong. Eventually James Maxwell and the Austrian physicist Ludwig Boltzmann understood. And what they understood is very beautiful, strange, and profound—and takes us into regions that are still largely unexplored.

What they came to understand is that a hot substance is not one that contains caloric fluid. A hot substance is a substance in which atoms move more quickly. Atoms and molecules, small clusters of atoms bound together, are always moving. They run, vibrate, bounce, and so on. Cold air is air in which atoms, or rather molecules, move more slowly. Hot air is air in which molecules move more rapidly. Beautifully simple. But it doesn't end there.

Heat, as we know, always moves from hot things to cold. A cold teaspoon placed in a cup of hot tea also becomes hot. If we don't dress accordingly on a freezing cold day, we quickly lose body heat and become cold. Why does heat go from hot things to cold things and not vice versa?

It is a crucial question because it relates to the nature of time. In every case in which heat exchange does not occur, or when the heat exchanged is negligible, we see that the future behaves exactly like the past. For example, for the motion of the planets of the solar system heat is almost irrelevant, and in fact this same motion could equally take place in reverse without any law of physics being infringed. As soon as there is heat, however, the future is different from the past. While there is no friction, for instance, a pendulum can swing forever. If we filmed it and ran the film in reverse, we would see movement that is completely possible. But if there is friction, then the pendulum heats its supports slightly, loses energy, and slows down. Friction produces heat. And immediately we are able to distinguish the future (toward which the pendulum slows) from the past. We have never seen a pendulum start swinging from a stationary position, with its movement initiated by the energy obtained by absorbing heat from its supports. The difference between past and future exists only when there is heat. The fundamental phenomenon that distinguishes the future from the past is the fact that heat passes from things that are hotter to things that are colder.

So, again, why, as time goes by, does heat pass from hot things to cold and not the other way around?

The reason was discovered by Boltzmann and is surprisingly simple: *it is sheer chance.*

Boltzmann's idea is subtle and brings into play the idea of probability. Heat does not move from hot things to cold things due to an absolute law: it does so only with a large degree of probability. The reason for this is that it is statistically more probable that a quickly moving atom of the hot substance collides with a cold one and leaves it a little of its energy, rather than vice versa. Energy is conserved in the collisions but tends to get distributed in more or less equal parts when there are many collisions. In this way the temperature of objects in contact with each other tends to equalize. It is not impossible for a hot body to become hotter through contact with a colder one: it is just extremely improbable.

This bringing of *probability* to the heart of physics, and using it to explain the bases of the dynamics of heat, was initially considered to be absurd. As frequently happens, no one took Boltzmann seriously. On September 5, 1906, in Duino, near Trieste, he committed suicide by hanging himself, never having witnessed the subsequent universal recognition of the validity of his ideas.

In the second lesson I related how quantum mechanics predicts that the movement of every minute thing occurs by chance. This puts probability into play as well. But the probability that Boltzmann considered, the probability at the roots of heat, has a different nature and is independent from quantum mechanics. The probability in play in the science of heat is in a certain sense tied to our *ignorance*.

I may not know something with certainty, but I can still assign a lesser or greater degree of probability to something. For instance, I don't know whether it will rain tomorrow here in Marseilles, or whether it will be sunny or will snow, but the probability that it will snow here tomorrow—in Marseilles, in August—is low. Similarly with regard to most physical objects: we know something but not everything about their state, and we can make predictions based only on probability. Think of a balloon filled with air. I can measure it: measure its shape, its volume, its pressure, its temperature . . . But the molecules of air inside the balloon are moving rapidly within it, and I do not know the exact position of each of them. This prevents me from predicting with precision how the balloon will behave. For instance, if I untie the knot that seals it and let it go, it will deflate noisily, rushing and colliding here and there in a way that is impossible for me to predict. Impossible, because I know only its shape, volume, pressure, and temperature. The bumping about here and there of the balloon depends on the detail of the position of the molecules inside it, which I don't know. Yet even if I can't predict everything exactly, I can predict the probability that one thing or another will happen. It will be very improbable, for instance, that the balloon will fly out of the window, circle the lighthouse down there in the distance, and then return to land on my hand, at the point where it was released. Some behavior is more probable, other behavior more improbable.

In this same sense, the probability that when molecules collide heat passes from the hotter bodies to those that are colder can be calculated, and turns out to be much greater than the probability of heat moving toward the hotter body.

The branch of science that clarifies these things is called "statistical physics," and one of its triumphs, beginning with Boltzmann, has been to understand the probabilistic nature of heat and temperature, that is to say, thermodynamics.

At first glance, the idea that our ignorance implies something about the behavior of the world seems irrational: the cold teaspoon heats up in hot tea and the balloon flies about when it is released regardless of what I know or don't know. What does what we know or don't know have to do with the laws that govern the world? The question is legitimate; the answer to it is subtle.

Teaspoon and balloon behave as they must, following the laws of physics in complete independence from what we know or don't know about them. The predictability or unpredictability of their behavior does not pertain to their precise condition; it pertains to the limited set of their properties with which we interact. *This* set of properties depends on *our* specific way of interacting with the teaspoon or balloon. Probability does not refer to the evolution of matter in itself. It relates to the evolution of those specific quantities we interact with. Once again, the profoundly relational nature of the concepts we use to organize the world emerges.

The cold teaspoon heats up in hot tea because tea and spoon interact with us through a limited number of variables among the innumerable variables that characterize their microstate. The value of *these* variables is not sufficient to predict future behavior exactly (witness the balloon) but is sufficient to predict with optimum probability that the spoon will heat up.

I hope not to have lost the reader's attention with these subtle distinctions . . .

Now, in the course of the twentieth century, thermodynamics (that is, the science of heat) and statistical mechanics (that is, the science of the probability of different motions) were extended to electromagnetic and quantum phenomena. Extension to include the gravitational field, however, has proved problematic. How the gravitational field behaves when it heats up is still an unsolved problem.

We know what happens to a heated electromagnetic field: in an oven, for instance, there is hot electromagnetic radiation, which cooks a pie, and we know how to describe this. The electromagnetic waves vibrate, randomly sharing energy, and we can imagine the whole as being like a gas of photons that move fast and randomly like the molecules in a hot balloon. But what is a hot *gravitational* field?

The gravitational field, as we saw in the first lesson, is space itself, in effect space-time. Therefore, when heat is diffused to the gravitational field, time and space themselves must vibrate . . . But we still don't know how to describe this well. We don't have the equations to describe the thermal vibrations of a hot space-time. What is a vibrating time?

Such issues lead us to the heart of the problem of time: what exactly is the *flow* of time?

The problem was already present in classical physics and was highlighted in the nineteenth and twentieth centuries by philosophers—but it becomes a great deal more acute in modern physics. Physics describes the world by means of formulas that tell how things vary as a function of "time." But we can write formulas that tell us how things vary in relation to their "position," or how the taste of a risotto varies as a function of the "variable quantity of butter." Time seems to "flow," whereas the quantity of butter or location in space does not "flow." Where does the difference come from?

Another way of posing the problem is to ask oneself: what is the "present"? We say that only the things of the present exist: the past no longer exists and the future doesn't exist yet. But in physics there is nothing that corresponds to the notion of the "now." Compare "now" with "here." "Here" designates the place where a speaker is: for two different people "here" points to two differ-ent places. Consequently "here" is a word the meaning of which depends on where it is spoken. The technical term for this kind of utterance is "indexical." "Now" also points to the instant in which the word is uttered and is also classed as "indexical." But no one would dream of saying that things "here" exist, whereas things that are not "here" do not exist. So then why do we say that things that are "now" exist and that everything else doesn't? Is the present something that is objective in the world, that "flows," and that makes things "exist" one after the other, or is it only subjective, like "here"?

This may seem like an abstruse mental problem. But modern physics has made it into a burning issue, since special relativity has shown that the notion of the "present" is also subjective. Physicists and philosophers have come to the conclusion that the idea of a present that is common to the whole universe is an illusion and that the universal "flow" of time is a generalization that doesn't work. When his great Italian friend Michele Besso died, Einstein wrote a moving letter to Michele's sister: "Michele has left this strange world a little before me. This means nothing. People like us, who believe in physics, know that the distinction made between past, present and future is nothing more than a persistent, stubborn illusion."

Illusion or not, what explains the fact that for us time "runs," "flows," "passes"? The passage of time is obvious to us all: our thoughts and our speech exist in time; the very structure of our language requires time—a thing "is" or "was" or "will be." It is possible to imagine a world without colors, without matter, even without space, but it's difficult to imagine one without time. The German philosopher Martin Heidegger emphasized our "dwelling in time." Is it possible that the flow of time that

Heidegger treats as primal is absent from descriptions of the world?

Some philosophers, the most devoted followers of Heidegger among them, conclude that physics is incapable of describing the most fundamental aspects of reality, and they dismiss it as a misleading form of knowledge. But many times in the past we have realized that it is our immediate intuitions that are imprecise: if we had kept to these we would still believe that Earth is flat and that it is orbited by the sun. Our intuitions have developed on the basis of our limited experience. When we look a little further ahead, we discover that the world is not as it appears to us: Earth is round, and in Cape Town their feet are up and their heads are down. To trust immediate intuitions rather than collective examination that is rational, careful, and intelligent is not wisdom: it is the presumption of an old man who refuses to believe that the great world outside his village is any different from the one that he has always known.

As vivid as it may appear to us, our experience of the passage of time does not need to reflect a fundamental aspect of reality. But if it is not fundamental, where does it come from, our vivid experience of the passage of time?

I think that the answer lies in the intimate connection between time and heat. There is a detectable difference between the past and the future only when there is the flow of heat. Heat is linked to probability; and probability in turn is linked to the fact that our interactions with the rest of the world do not register the fine details of reality. The flow of time emerges thus from physics, but not in the context of an exact description of things as they are. It emerges, rather, in the context of statistics and of thermodynamics. This may hold the key to the enigma of time. The "present" does not exist in an objective sense any more than "here" exists objectively, but the microscopic interactions within the world prompt the emergence of temporal phenomena within a system (for instance, ourselves) that interacts only through the medium of a myriad of variables.

Our memory and our consciousness are built on these statistical phenomena. For a hypothetically supersensible being, there would be no "flowing" of time: the universe would be a single block of past, present, and future. But due to the limitations of our consciousness we perceive only a blurred vision of the world and live in time. Borrowing words from my Italian editor, "what's non-apparent is much vaster than what's apparent." From this limited, blurred focus we get our perception of the passage of time. Is that clear? No, it isn't. There is so much still to be understood.

Time sits at the center of the tangle of problems raised by the intersection of gravity, quantum mechanics, and thermodynamics. A tangle of problems where we are still in the dark. If there is something that we are perhaps beginning to understand about quantum gravity that combines two of the three pieces of the puzzle, we do not yet have a theory capable of drawing together all three pieces of our fundamental knowledge of the world.

A small clue toward the solution comes from a calculation completed by Stephen Hawking, the physicist famous for having continued to produce outstanding physics despite a medical condition that keeps him confined to a wheelchair and prevents him from speaking without a mechanical aid.

Using quantum mechanics, Hawking successfully demonstrated that black holes are always "hot." They emit heat like a stove. It's the first concrete indication on the nature of "hot space." No one has ever observed this heat because it is faint in the actual black holes that have been observed so far—but Hawking's calculation is convincing, it has been repeated in different ways, and the reality of the heat of black holes is generally accepted.

The heat of black holes is a quantum effect upon an object, the black hole, which is gravitational in nature. It is the individual quanta of space, the elementary grains of space, the vibrating "molecules," that heat the surface of black holes and generate black hole heat. This phenomenon involves all three sides of the problem: quantum mechanics, general relativity, and thermal science. The heat of black holes is like the Rosetta stone of physics, written in a combination of three languages— quantum, gravitational, and thermodynamic—still awaiting decipherment in order to reveal the true nature of time.

Ourselves

After having journeyed so far, from the structure of deep space to the margins of the known cosmos, I would like to return, before closing this series of lessons, to the subject of ourselves.

What role do we have as human beings who perceive, make decisions, laugh, and cry, in this great fresco of the world as depicted by contemporary physics? If the world is a swarm of ephemeral quanta of space and matter, a great jigsaw puzzle of space and elementary particles, then what are we? Do we also consist only of quanta and particles? If so, then from where do we get that sense of individual existence and unique selfhood to which we can all testify? And what then are our values, our dreams, our emotions, our individual knowledge? What are we, in this boundless and glowing world?

I cannot even imagine attempting to really answer such a question in these simple pages. It's a tough question. In the big picture of contemporary science, there are many things that we do not understand, and one of the things that we understand least about is ourselves. But to avoid this question or to ignore it would be, I think, to overlook something essential. I've set out to show how the world looks in the light of science, and we are a part of that world too.

"We," human beings, are first and foremost the subjects who do the observing of this world, the collective makers of the photograph of reality that I have tried to compose. We are nodes in a network of exchanges (of which this present book is an example) through which we pass images, tools, information, and knowledge.

But we are also an integral part of the world that we perceive; we are not external observers. We are situated within it. Our view of it is from within its midst. We are made up of the same atoms and the same light signals as are exchanged between pine trees in the mountains and stars in the galaxies.

As our knowledge has grown, we have learned that

our being is only a part of the universe, and a small part at that.

This has been increasingly apparent for centuries, but especially so during the last century. We believed that we were on a planet at the center of the universe, and we are not. We thought that we existed as unique beings, a race apart from the family of animals and plants, and discovered that we are descendants of the same parents as every living thing around us. We have great-grandparents in common with butterflies and larches. We are like an only child who in growing up realizes that the world does not revolve only around himself, as he thought when little. He must learn to be one among others. Mirrored by others, and by other things, we learn who we are.

During the great period of German idealism, Schelling could think that man represented the summit of nature, the highest point where reality becomes conscious of itself. Today, from the point of view provided by our current knowledge of the natural world, this idea raises a smile. If we are special, we are only special in the way that everyone feels themselves to be, like every mother is for her child. Certainly not for the rest of nature.

Within the immense ocean of galaxies and stars we

are in a remote corner; amid the infinite arabesques of forms that constitute reality, we are merely a flourish among innumerably many such flourishes.

The images that we construct of the universe live within us, in the space of our thoughts. Between these images—between what we can reconstruct and understand with our limited means—and the reality of which we are part, there exist countless filters: our ignorance, the limitations of our senses and of our intelligence. The very same conditions that our nature as subjects, and particular subjects, imposes upon experience.

These conditions, nevertheless, are not, as Kant imagined, universal—deducing from this (with obvious error) that the nature of Euclidian space and even of Newtonian mechanics must therefore be true a priori. They are a posteriori to the mental evolution of our species and are in continuous evolution. We not only learn, but we also learn to gradually change our conceptual framework and to adapt it to what we learn. And what we are learning to recognize, albeit slowly and hesitantly, is the nature of the real world of which we are part. The images that we construct of the universe may live inside us, in conceptual space, but they also describe more or less well the real world to which we belong. We follow leads in order to better describe this world.

When we talk about the big bang or the fabric of space, what we are doing is not a continuation of the free and fantastic stories that humans have told nightly around campfires for hundreds of thousands of years. It is the continuation of something else: of the gaze of those same men in the first light of day looking at tracks left by antelope in the dust of the savannah—scrutinizing and deducting from the details of reality in order to pursue something that we can't see directly but can follow the traces of. In the awareness that we can always be wrong, and therefore ready at any moment to change direction if a new track appears; but knowing also that if we are good enough we will get it right and will find what we are seeking. This is the nature of science.

The confusion between these two diverse human activities—inventing stories and following traces in order to find something—is the origin of the incomprehension and distrust of science shown by a significant part of our contemporary culture. The separation is a subtle one: the antelope hunted at dawn is not far removed from the antelope deity in that night's storytelling.

The border is porous. Myths nourish science, and science nourishes myth. But the value of knowledge remains. If we find the antelope, we can eat.

Our knowledge consequently reflects the world. It

does this more or less well, but it reflects the world we inhabit. This communication between ourselves and the world is not what distinguishes us from the rest of nature. All things are continually interacting with one another, and in doing so each bears the traces of that with which it has interacted: and in this sense all things continuously exchange information about one another.

The information that one physical system has about another has nothing mental or subjective about it: it's only the connection that physics determines between the state of something and the state of something else. A raindrop contains information on the presence of a cloud in the sky, a ray of light contains information on the color of the substance from which it came, a clock has information on the time of day, the wind carries information about an approaching storm, a cold virus has information of the vulnerability of my nose, the DNA in our cells contains all the information in our genetic code (on what makes me resemble my father), and my brain teems with information accumulated from my experience. The primal substance of our thoughts is an extremely rich gathering of information that's accumulated, exchanged, and continually elaborated.

Even the thermostat on my central heating system

"senses" and "knows" the ambient temperature in my home, has information on it, and turns off when it is warm enough. So what then is the difference between the thermostat's and my own "sensing" and "knowing" that it's warm and deciding freely to turn off the heating or not—and "knowing" that I exist? How can the continuous exchange of information in nature produce *us* and our thoughts?

The problem is wide-open, with numerous fine solutions currently under discussion. This, I believe, is one of the most interesting frontiers of science, where major progress is about to be made. Today new tools allow us to observe the activity of the brain in action and to map its highly intricate networks with impressive precision. As recently as 2014 the news was announced that the first complete (mesoscopic) detailed mapping of the brain structure of a mammal had been achieved. Specific ideas on how the mathematical form of the structures can correspond to the subjective experience of consciousness are currently being discussed, not only by philosophers but also by neuroscientists.

An intriguing one, for instance, is the mathematical theory being developed by Giulio Tononi—an Italian scientist working in the United States. It's called "integrated information theory" and is an attempt to characterize quantitatively the structure that a system must have in order to be conscious: a way, for example, of describing what actually changes on the physical plane between when we are awake (conscious) and when we are asleep but not dreaming (unconscious). It's still at the developmental phase. We still have no convincing and established solution to the problem of how our consciousness is formed. But it seems to me that the fog is beginning to clear.

There is one issue in particular regarding ourselves that often leaves us perplexed: what does it mean, our being free to make decisions, if our behavior does nothing but follow the predetermined laws of nature? Is there not perhaps a contradiction between our feeling of freedom and the rigor, as we now understand it, with which things operate in the world? Is there perhaps something in us that escapes the regularity of nature and allows us to twist and deviate from it through the power of our freedom to think?

Well, no, there is nothing about us that can escape the norms of nature. If something in us could infringe the laws of nature, we would have discovered it by now. There is nothing in us in violation of the natural behavior of things. The whole of modern science—from physics to chemistry, and from biology to neuroscience—does nothing but confirm this observation.

The solution to the confusion lies elsewhere. When we say that we are free, and it's true that we can be, this means that how we behave is determined by what happens within us, within the brain, and not by external factors. To be free doesn't mean that our behavior is not determined by the laws of nature. It means that it is determined by the laws of nature acting in our brains.

Our free decisions are freely determined by the results of the rich and fleeting interactions among the billion neurons in our brain: they are free to the extent that the interaction of these neurons allows and determines. Does this mean that when I make a decision it's "I" who decides? Yes, of course, because it would be absurd to ask whether "I" can do something different from what the whole complex of my neurons has decided: the two things, as the Dutch philosopher Baruch Spinoza understood with marvelous lucidity in the seventeenth century, are the same.

There is not an "I" *and* "the neurons in my brain." They are the same thing. An individual is a process: complex, tightly integrated.

When we say that human behavior is unpredictable, we are right because it is too complex to be predicted, especially by ourselves. Our intense sensation of interior liberty, as Spinoza acutely saw, comes from the fact that the ideas and images that we have of ourselves are much cruder and sketchier than the detailed complexity of what is happening within us. We are the source of amazement in our own eyes.

We have a hundred billion neurons in our brains, as many as there are stars in a galaxy, with an even more astronomical number of links and potential combinations through which they can interact. We are not conscious of all of this. "We" are the process formed by this entire intricacy, not just by the little of it of which we are conscious.

The "I" who decides is that same "I" that is formed (in a way that is still certainly not completely clear, but that we have begun to glimpse) from reflections upon itself, through self-representations in the world, from understanding itself as a variable point of view placed in the context of the world, from that impressive structure that processes information and constructs representations that is our brain. When we have the feeling that "it is I" who decides, we couldn't be more correct. Who else?

I am, as Spinoza maintained, my body and what happens in my brain and heart, with their immense and, for me, inextricable complexity.

The scientific picture of the world that I have related in these pages is not, then, at odds with our sense of ourselves. It is not at odds with our thinking in moral and psychological terms, or with our emotions and feelings. The world is complex, and we capture it with different languages, each appropriate to the process that we are describing. Every complex process can be addressed and understood in different languages and at different levels. These diverse languages intersect, intertwine, and reciprocally enhance one another, like the processes themselves. The study of our psychology becomes more sophisticated through our understanding of the biochemistry of the brain. The study of theoretical physics is nourished by the passions and emotions that animate our lives.

Our moral values, our emotions, our loves are no less real for being part of nature, for being shared with the animal world, or for being determined by the evolution that our species has undergone over millions of years. Rather, they are more valuable as a result of this: they are real. They are the complex reality of which we are made. Our reality is tears and laughter, gratitude and altruism, loyalty and betrayal, the past that haunts us and serenity. Our reality is made up of our societies, of the emotion inspired by music, of the rich intertwined networks of the common knowledge that we have constructed together. All of this is part of the self-same "nature" that we are describing. We are an integral part of nature; we *are* nature, in one of its innumerable and infinitely variable expressions. This is what we have learned from our ever-increasing knowledge of the things of this world.

That which makes us specifically human does not signify our separation from nature; it is part of that self-same nature. It's a form that nature has taken here on our planet, in the infinite play of its combinations, through the reciprocal influencing and exchanging of correlations and information among its parts. Who knows how many and which other extraordinary complexities exist, in forms perhaps impossible for us to imagine, in the endless spaces of the cosmos? There is so much space up there that it is childish to think that in a peripheral corner of an ordinary galaxy there should be something uniquely special. Life on Earth gives only a small taste of what can happen in the universe. Our very soul itself is only one such small example.

We are a species that is naturally moved by curiosity, the only one left of a group of species (the genus *Homo*) made up of a dozen equally curious species. The other species in the group have already become extinct—some, like the Neanderthals, quite recently, roughly thirty thousand years ago. It is a group of species that evolved in Africa, akin to the hierarchical and quarrelsome chimpanzees—and even more closely akin to the bonobos, the small, peaceful, cheerfully egalitarian, and promiscuous type of chimps. A group of species that repeatedly went out of Africa in order to explore new worlds, and went far: as far, eventually, as Patagonia—and as far, eventually, as the moon.

It is not against nature to be curious: it is in our nature to be so.

One hundred thousand years ago our species left Africa, compelled perhaps by precisely this curiosity, learning to look ever farther afield. Flying over Africa by night, I wondered if one of these distant ancestors setting out toward the wide-open spaces of the North could have looked up into the sky and imagined a distant descendant flying up there, pondering on the nature of things, and still driven by his very same curiosity.

I believe that our species will not last long. It does

not seem to be made of the stuff that has allowed the turtle, for example, to continue to exist more or less unchanged for hundreds of millions of years, for hundreds of times longer, that is, than we have even been in existence. We belong to a short-lived genus of species. All of our cousins are already extinct. What's more, we do damage. The brutal climate and environmental changes that we have triggered are unlikely to spare us. For Earth they may turn out to be a small irrelevant blip, but I do not think that we will outlast them unscathed—especially since public and political opinion prefers to ignore the dangers that we are running, hiding our heads in the sand. We are perhaps the only species on Earth to be conscious of the inevitability of our individual mortality. I fear that soon we shall also have to become the only species that will knowingly watch the coming of its own collective demise, or at least the demise of its civilization.

As we know more or less well how to deal with our individual mortality, so we will deal with the collapse of our civilization. It is not so different. And it's certainly not the first time that this will have happened. The Maya and Cretans, among many others, already experienced this. We are born and die as the stars are born and die, both individually and collectively. This is our reality. Life is precious to us because it is ephemeral. And as Lucretius wrote: "our appetite for life is voracious, our thirst for life insatiable" (*De rerum natura*, bk. III, line 1084). But immersed in this nature that made us and that directs us, we are not homeless beings suspended between two worlds, parts *of* but only partly belonging *to* nature, with a longing for something else. No: we are home.

Nature is our home, and in nature we are *at* home.

This strange, multicolored, and astonishing world that we explore—where space is granular, time does not exist, and things are nowhere—is not something that estranges us from our true selves, for this is only what our natural curiosity reveals to us about the place of our dwelling. About the stuff of which we ourselves are made. We are made of the same stardust of which all things are made, and when we are immersed in suffering or when we are experiencing intense joy, we are being nothing other than what we can't help but be: a part of our world.

Lucretius expresses this, wonderfully:

> . . . we are all born from the same celestial
>   seed;
> all of us have the same father,
> from which the earth, the mother who
>   feeds us,
> receives clear drops of rain,
> producing from them bright wheat
> and lush trees,
> and the human race,
> and the species of beasts,
> offering up the foods with which all bodies are
>   nourished,
> to lead a sweet life
> and generate offspring . . .

(De rerum natura, bk. II, lines 991–97)

It is part of our nature to love and to be honest. It is part of our nature to long to know more and to continue to learn. Our knowledge of the world continues to grow.

There are frontiers where we are learning, and our

desire for knowledge burns. They are in the most minute reaches of the fabric of space, at the origins of the cosmos, in the nature of time, in the phenomenon of black holes, and in the workings of our own thought processes. Here, on the edge of what we know, in contact with the ocean of the unknown, shines the mystery and the beauty of the world. And it's breathtaking.

INDEX